How to use this book

INSTRUCTION
What your child needs to do for the activity.

TITLE
The page title describes the skill your child will learn in these pages.

FUN ILLUSTRATIONS
Specially drawn illustrations which are fun, interesting and drawn at the right pedagogical level for your child.

COLOURFUL BORDERS
The page borders make each page as attractive as possible to stimulate your child.

EXAMPLE
The first one is done for you so you can show your child exactly what to do.

LOTS OF PRACTICE
Two pages where your child can practise and repeat the same skill to master it.

STICKERS
Place a sticker on each page as your child finishes.

WHO'S HIDING?
In each book, a little creature appears in the border of every double page so your child can have fun trying to find it.

EXTRA ACTIVITIES
Extra activities you might want to do with your child to further reinforce the skill or simply make it more enjoyable.

Step-by-step learning

STEP ONE **Read** out the title of the activity page to your child.

STEP TWO **Explain** the skill and show your child the example already done. **Make sure** they understand what to do. Your child will then have at least two pages to practise that same skill.

STEP THREE **Help** your child put a **sticker** on the bottom of each page as they complete it.

Remember to be patient, encouraging and positive with your child, even when minor mistakes are made!

How to hold a pencil

It is important that you help your child hold their crayon or pencil in the correct way, as shown here, to ensure your child develops the right technique early on.

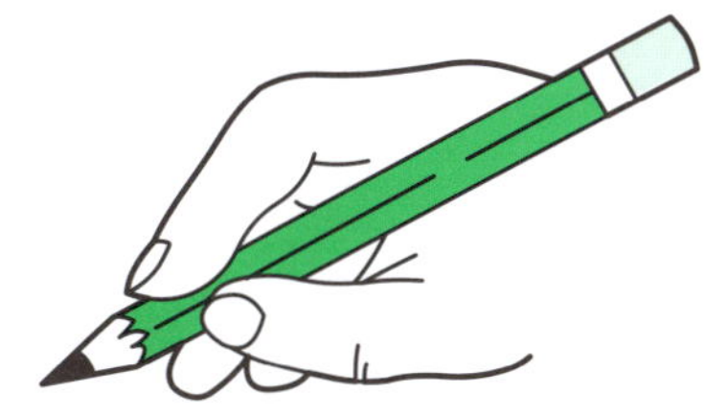

Matching one-for-one

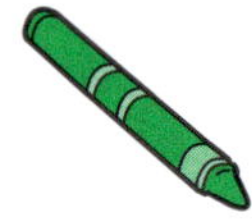

Join each hat to a face. Then join each piece of cheese to a mouse.

1. Can your child count how many hats are on the page?

Place a sticker here.

2. Can your child count how many pieces of cheese are on the page?

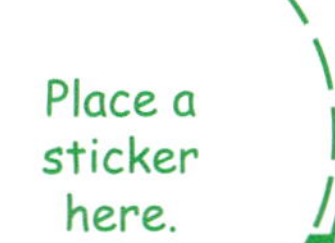

Making patterns of 1

Pick two different coloured crayons. Colour the first picture in one colour, then the second picture in the next colour. Then make the same pattern again until you finish the row.

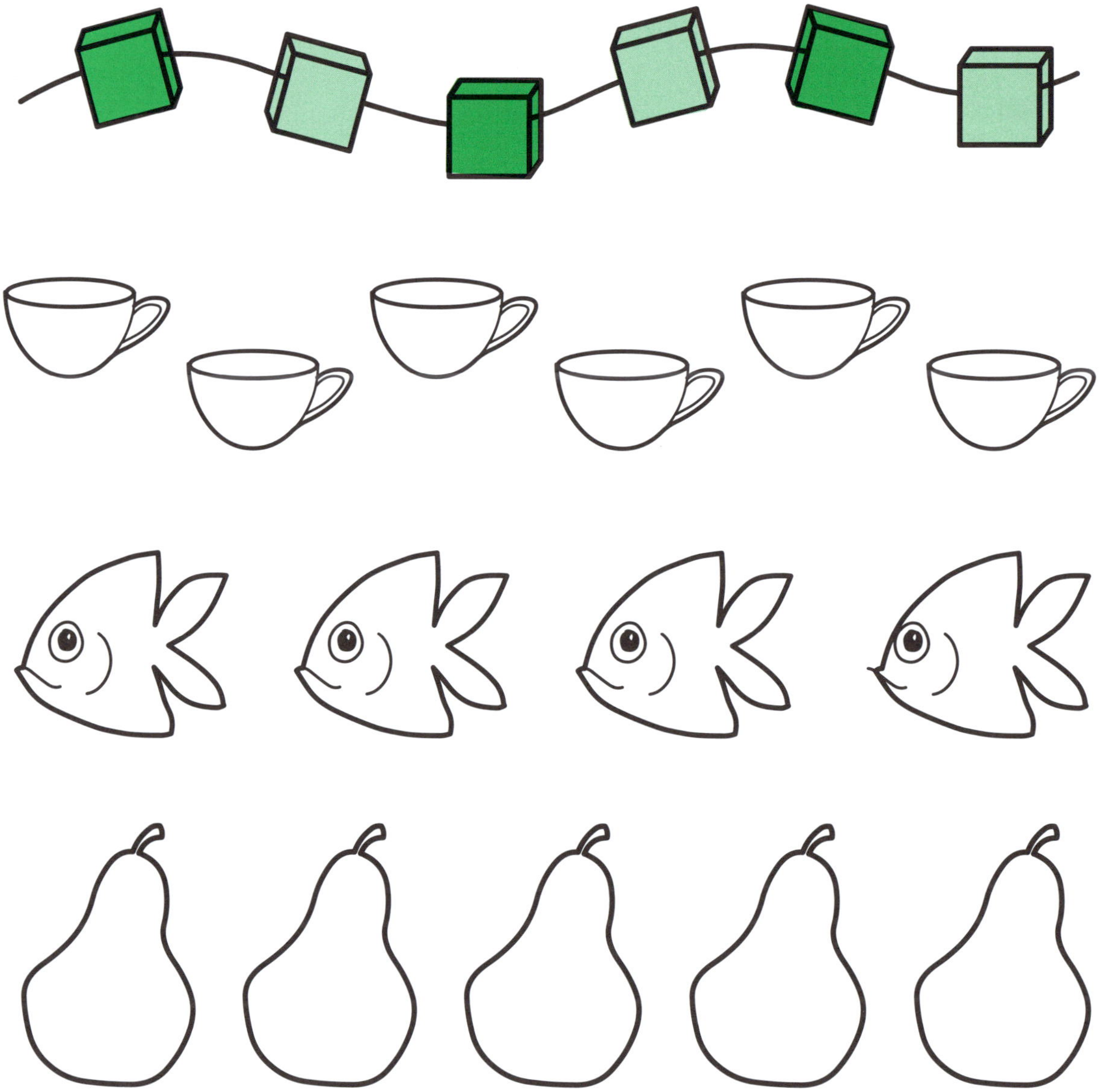

1. Ask your child to count the pictures in each row.

Place a sticker here.

2. Ask your child to find the row which has the *fewest* pictures in it.

Place a sticker here.

Colouring groups of 1

How many different animals can you see? Colour one of each type in blue, one in red and one in green.

(For example, one fish will be blue, one will be red and one will be green.)

1. Ask your child to count how many animals are green.

Place a sticker here.

2. Ask your child to count how many animals are blue.

Place a sticker here.

Writing the number 1

How do you draw the number 1? Start at the top and drag your crayon or pencil down all the dotted lines on the page. Then draw 1 spot on the dog.

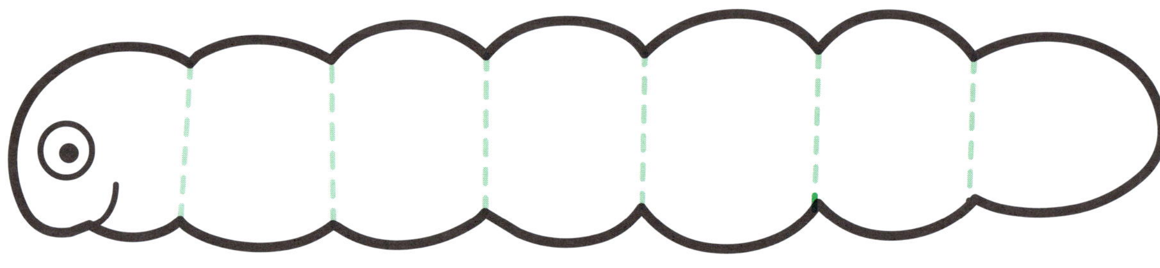

1. How many number 1s can your child see on this page?

Place a sticker here.

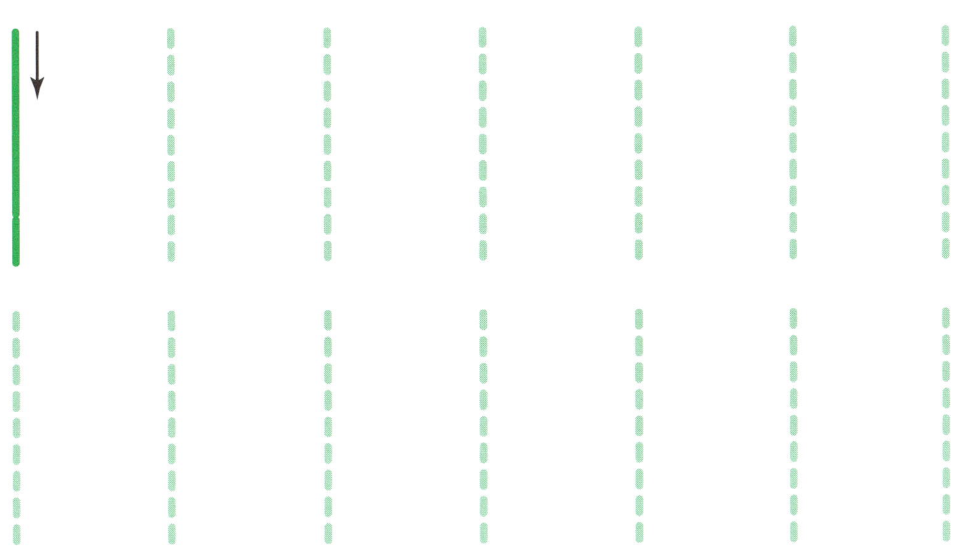

1 spot

2. A good way to start writing numbers is to have your child practise drawing each new number in the air before writing it.

Place a sticker here.

Making patterns of 2

Pick two different coloured crayons. Colour the first 2 pictures in one colour, then the next 2 pictures in the other colour. Then make the same pattern again until you finish the row.

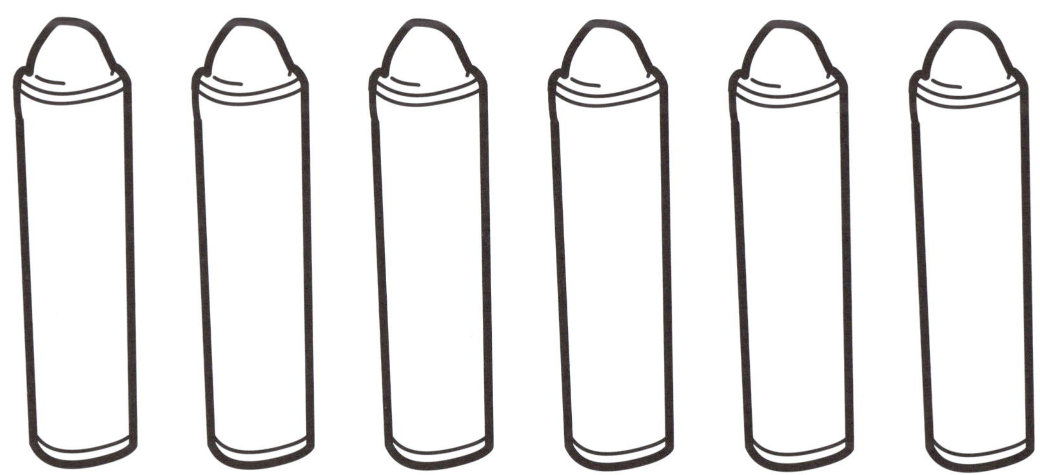

1. Ask your child to count the objects in each row.

Place a sticker here.

2. Ask your child to find the row on this page which has the *most* pictures in it.

Place a sticker here.

Drawing groups of 2

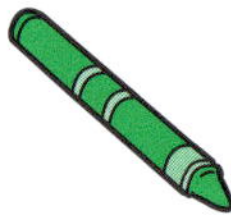

Draw matching pictures so there are 2 things.

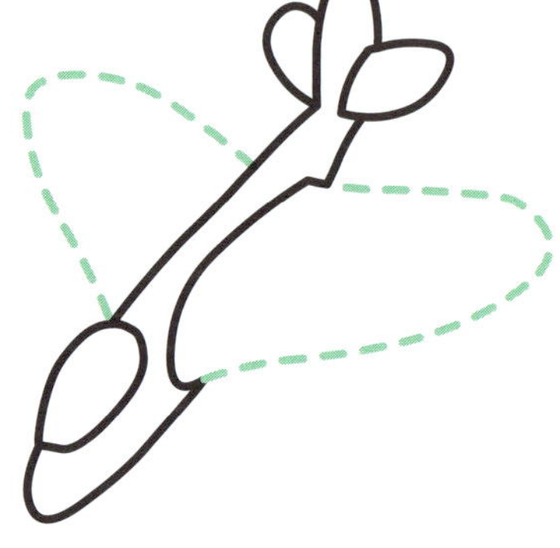

2 wings

2 wheels

1. See if your child can find other groups of 2 around you in the home. Are there 2 taps in the sink, or 2 matching doors on a cupboard?

Place a sticker here.

2 legs

2 balls

2. Ask your child to colour in each group of 2 in the same coloured crayon.

Place a sticker here.

Writing the number 2

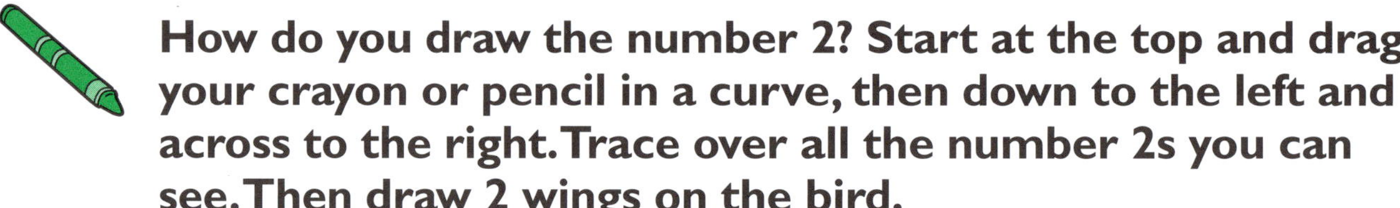

How do you draw the number 2? Start at the top and drag your crayon or pencil in a curve, then down to the left and across to the right. Trace over all the number 2s you can see. Then draw 2 wings on the bird.

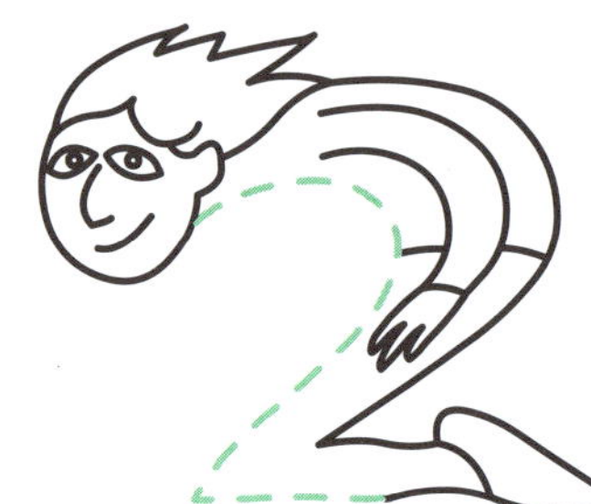

1. How many number 2s can your child see on this page?

Place a sticker here.

2 2 2 2

2 2 2 2

2 wings

2. See how many number 2s your child can find around you in the home — the bigger the better!

Place a sticker here.

Making patterns of 3

Pick two different coloured crayons. Colour the first 3 pictures in one colour, then the next 3 pictures in the other colour. Then make the same pattern again until you finish the row.

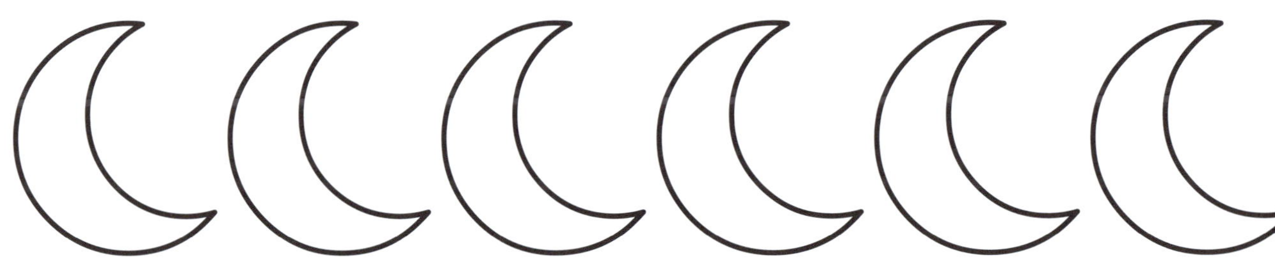

1. Ask your child to count the pictures in each group.

2. Ask your child to find the group on these pages which has the *most* pictures in it.

Place a sticker here.

Drawing groups of 3

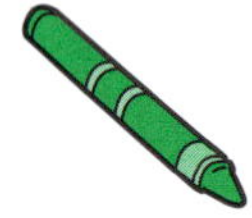

Draw 3 things in each group.

3 eggs

3 balls

1. See if your child can find other groups of 3 around you in the home. Are there 3 cups that look the same? Or 3 plants with the same pot?

Place a sticker here.

3 glasses

3 candles

3 faces

2. Ask your child to colour in each group of 3 in the same coloured crayon.

Place a sticker here.

Writing the number 3

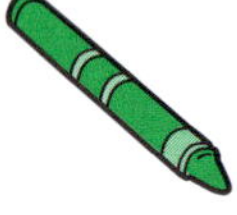

How do you draw the number 3? Start at the top and drag your crayon or pencil in two curves, down and down again. Trace over all the number 3s you can see. Then draw 3 feathers on the rooster.

1. How many number 3s can your child see on this page?

Place a sticker here.

3 feathers

2. See how many number 3s your child can find around you in the home.

Place a sticker here.

Finding the matching number

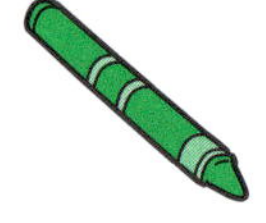

How many things can you see in each box? Circle the matching number.

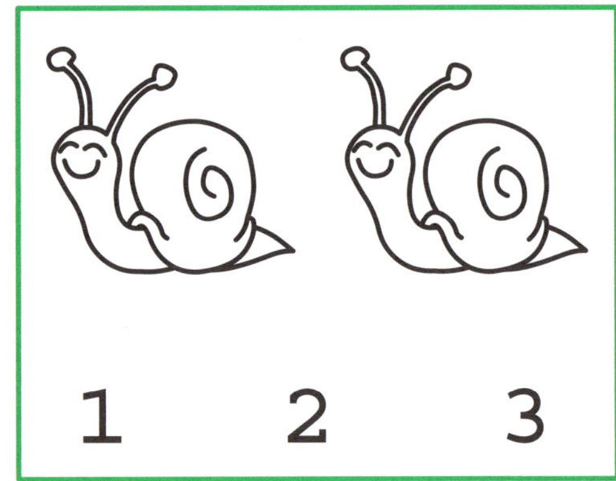

1. How far can your child count past 3? Can they count how many animals there are on this page?

Place a sticker here.

from 1-3

1 2 3

1 2 3

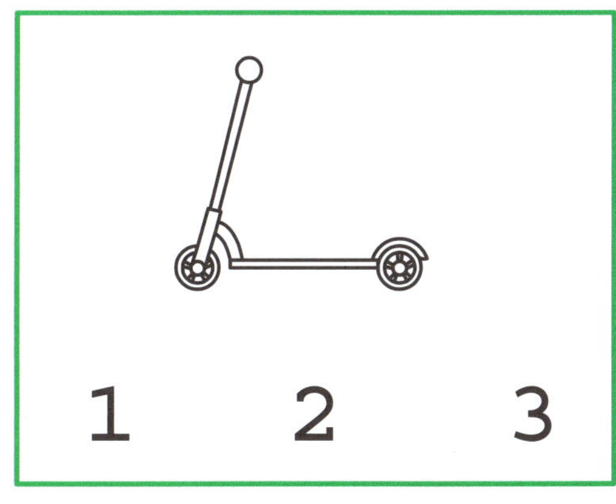
1 2 3

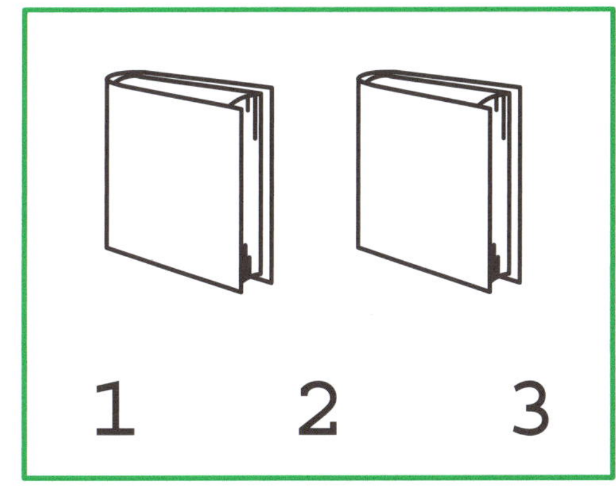
1 2 3

1 2 3

1 2 3

2. Can your child count how many toys there are on this page?

Place a sticker here.

Writing the matching number

Count the pictures in each group. Write the matching number as large as you can on the line beside them.

1. Can your child name all the fruits on this page?

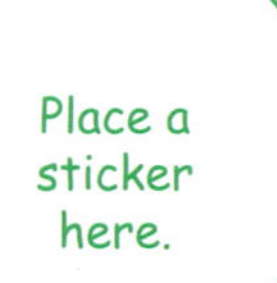

2. Ask your child where they might find all these animals.

Place a sticker here.

Making patterns of 4

Pick two different coloured crayons. Colour the first 4 shapes in one colour, then the next 4 shapes in the other colour. Then make the same pattern again until you finish the group.

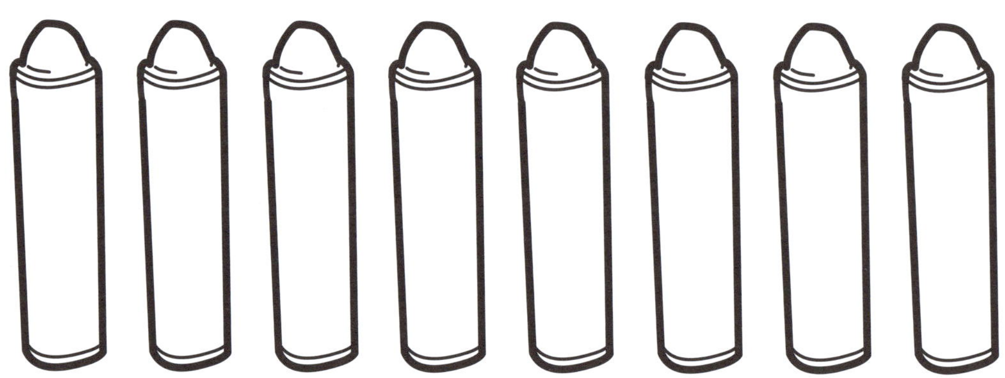

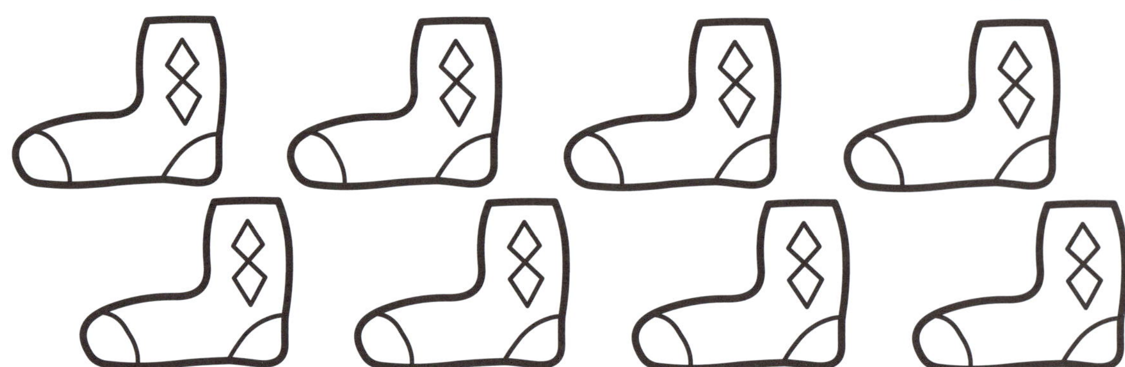

1. Ask your child to count the pictures in each group.

Place a sticker here.

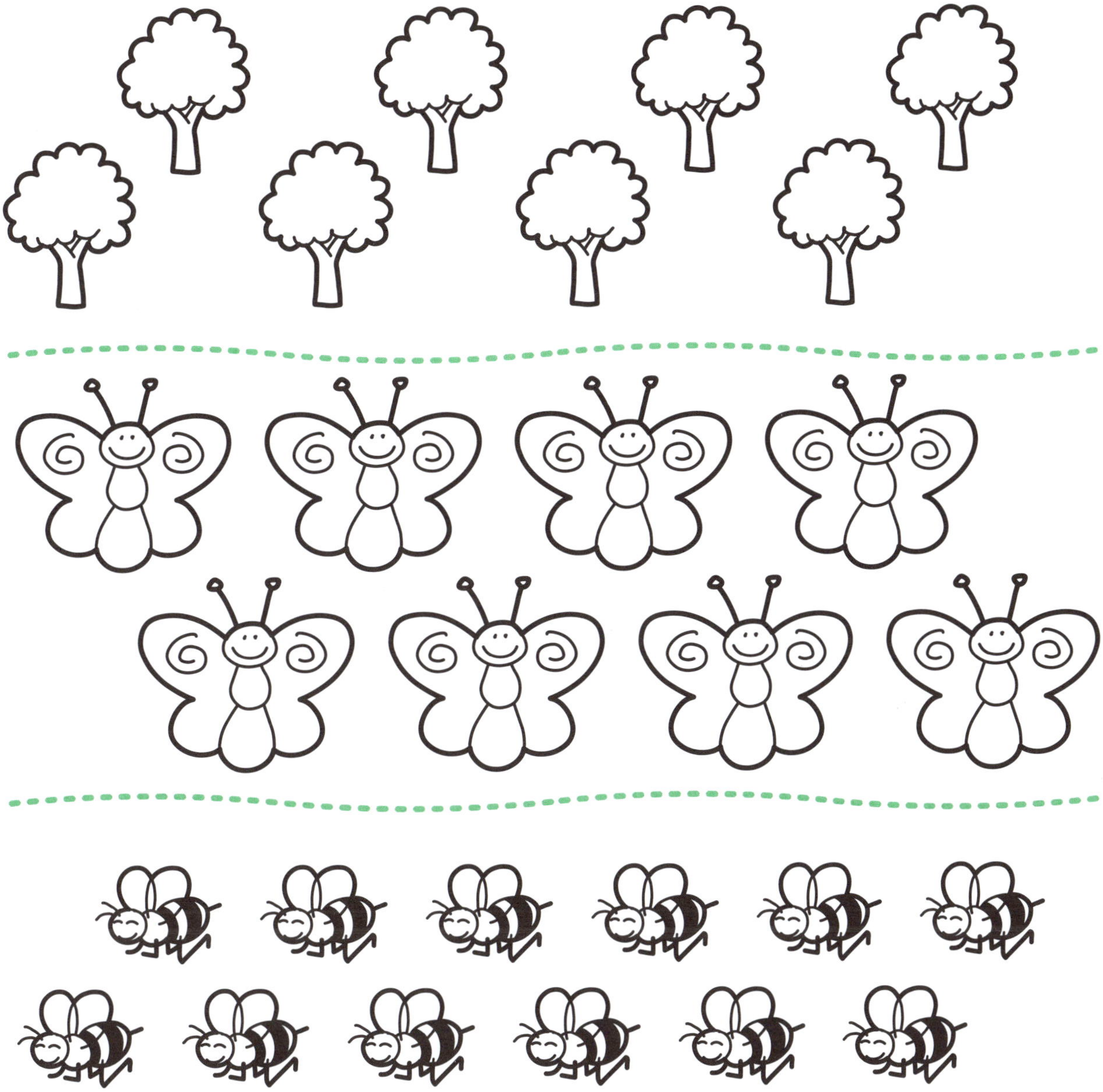

2. Ask your child if there are more trees or more bees on this page..

Place a sticker here.

Drawing groups of 4

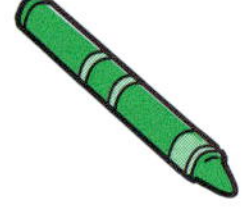

Draw matching pictures so there are 4 things in each group.

4 bones

4 balls

4 candles

1. Ask your child to find other groups of 4 around you in the home. Are there 4 plates that look the same? Or 4 legs on the table?

Place a sticker here.

4 apples

4 spots

2. Ask your child to colour in each group of 4 in the same coloured crayon.

Place a sticker here.

Writing the number 4

How do you draw the number 4? Start at the top and drag your crayon or pencil to the left and then straight across. Then draw a short line straight down. Trace over all the number 4s you can see. Then draw 4 stripes on each ball.

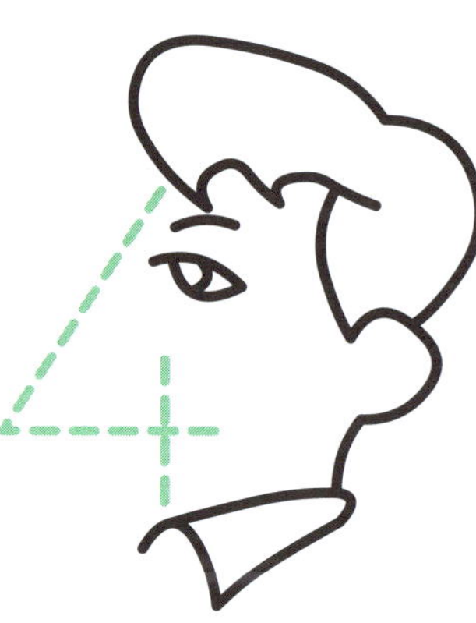

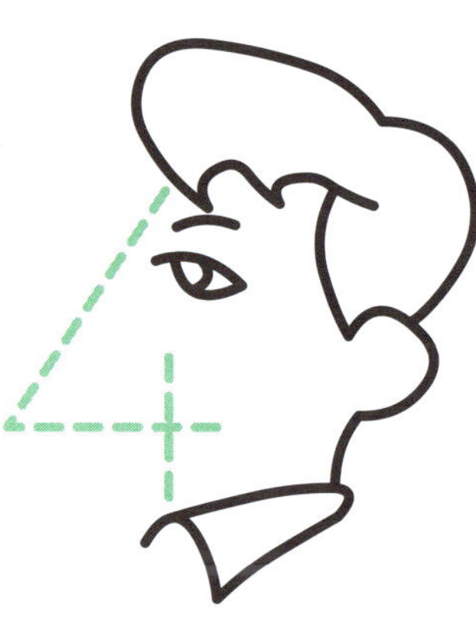

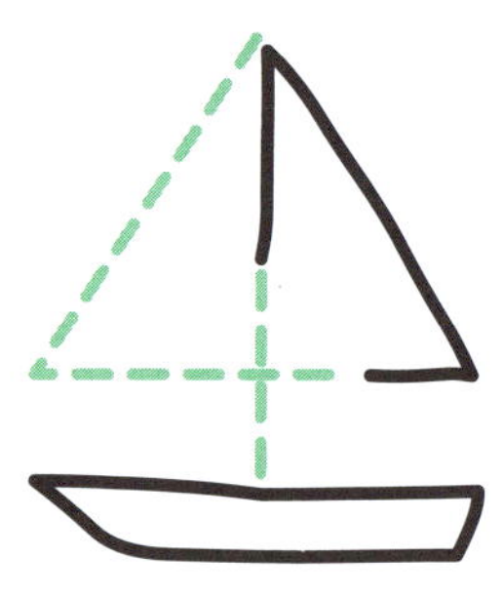

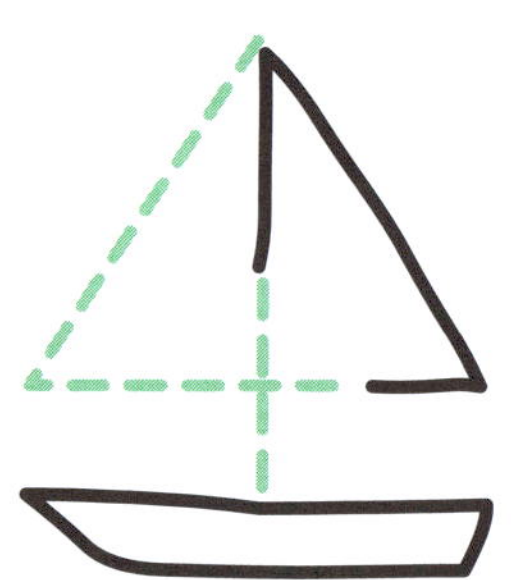

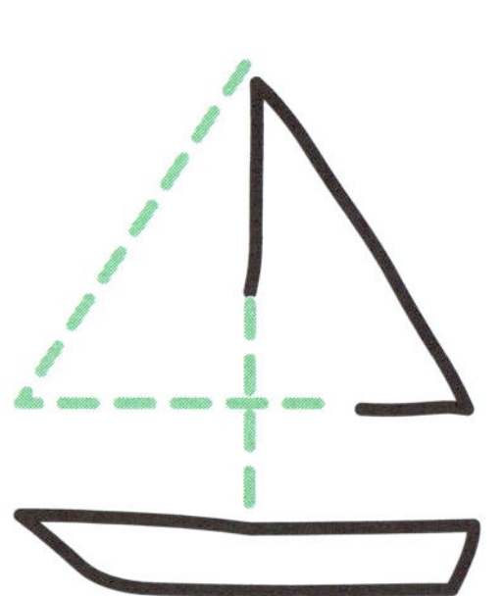

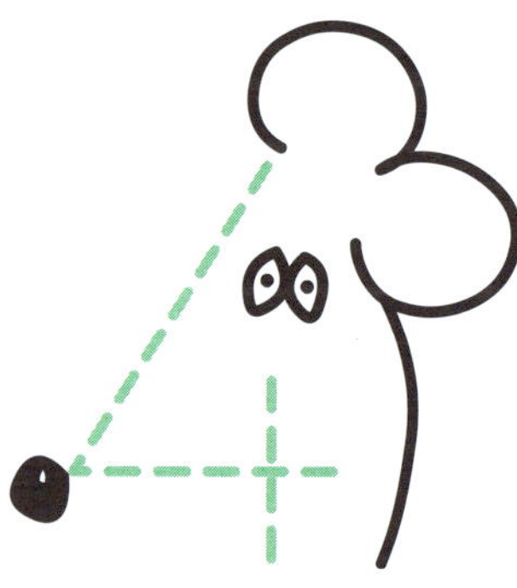

1. How many number 4s can your child see on this page?

Place a sticker here.

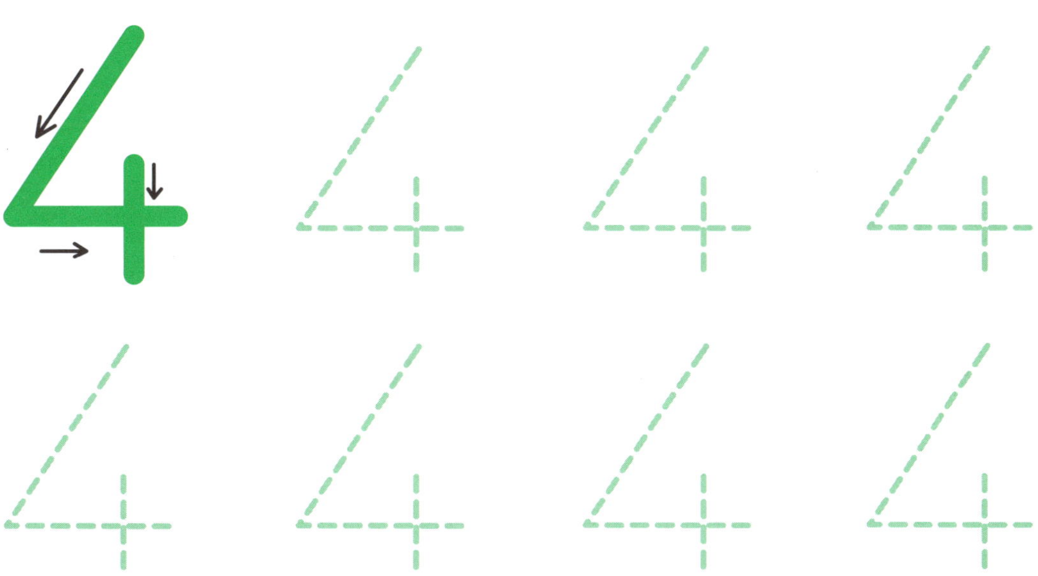

4 stripes

2. See how many number 4s your child can find around you when you next go out for a walk. Are there any on numberplates? On houses? On signs?

Place a sticker here.

Making patterns of 5

Pick two different coloured crayons. Colour the first 5 shapes in one colour, then the next 5 shapes in the other colour. Then make the same pattern again until you finish the group.

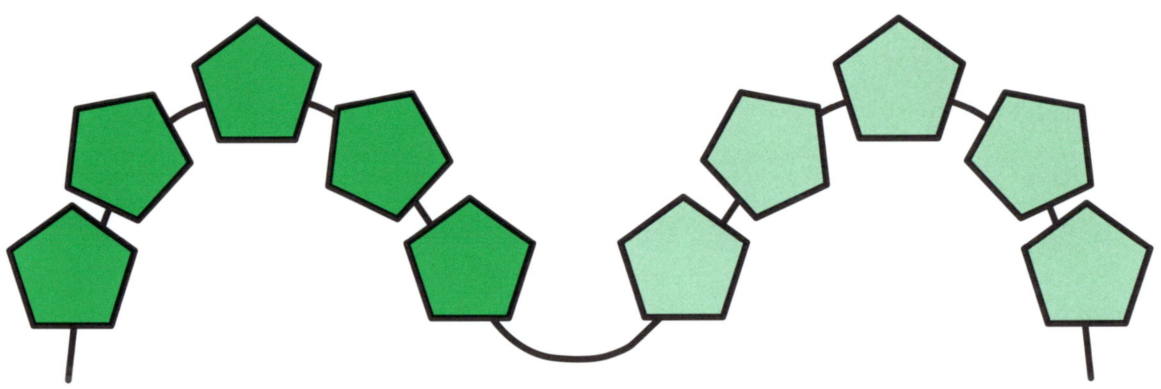

1. Ask your child to count the pictures in each group.

Place a sticker here.

2. Ask your child to find the group which has the *most* pictures in it.

Place a sticker here.

Drawing groups of 5

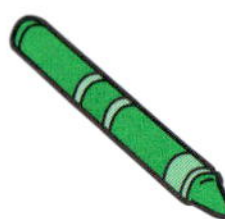

Draw matching pictures so there are 5 things in each group

5 lilypads

5 teeth

1. See if your child can see any other groups of 5 around you in the home. Are there 5 plastic containers that look the same? Or 5 bottles of drink?

Place a sticker here.

5 rungs

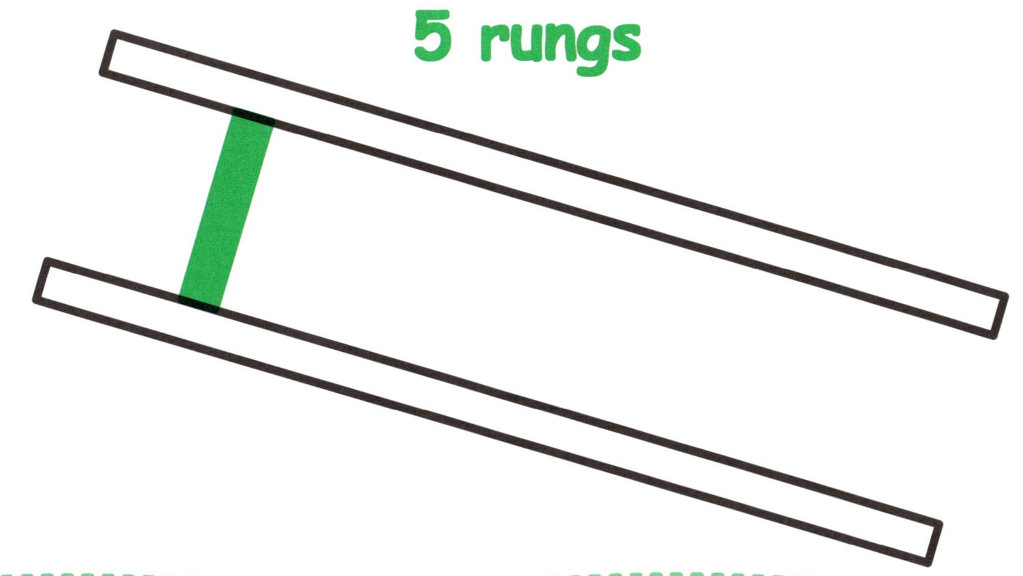

5 stripes

2. Ask your child to collect 5 objects from their bedroom.

Place a sticker here.

Writing the number 5

How do you draw the number 5? Start at the top left. Drag your crayon or pencil down, the curve it around. Now draw in the top line from left to right. Trace over all the number 5s you can see. Then draw 5 spots on each frog.

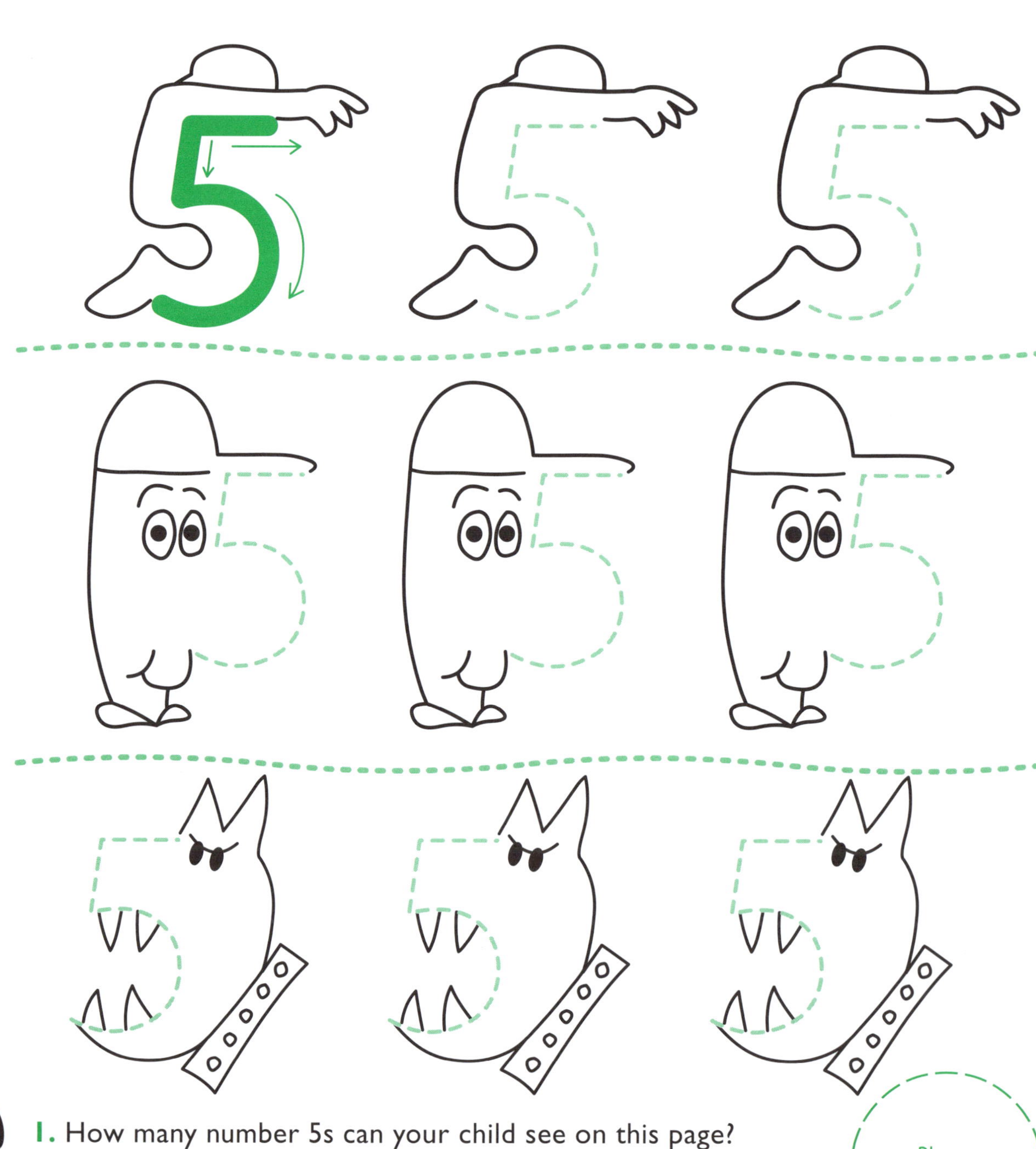

1. How many number 5s can your child see on this page?

Place a sticker here.

5 spots

2. See how many number 5s your child can find around you when you next go to the shop. Encourage your child to count 5 benches, or 5 shops in a row.

Place a sticker here.

Joining the matching sets

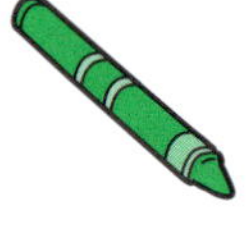

How many animals are in each box? Draw a line between the boxes which have the same number of animals inside them.

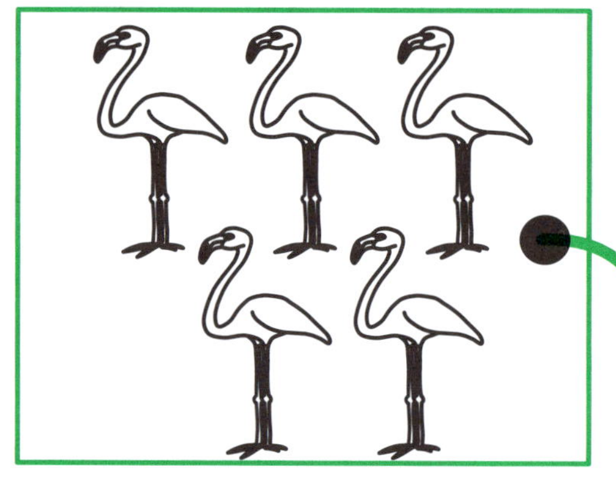

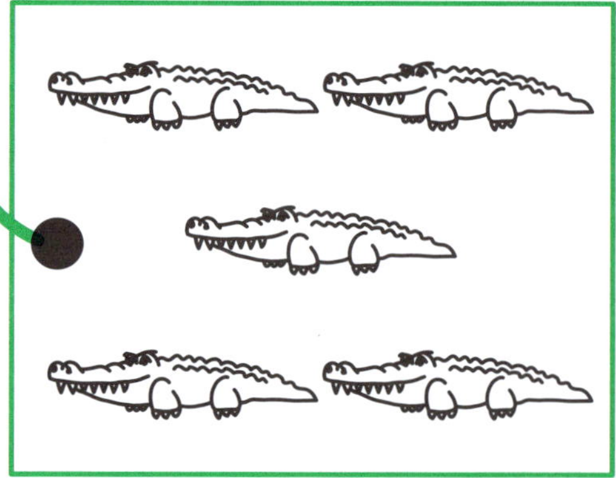

1. Ask your child if there are more monkeys than bears on the page. Then count them to check.

Place a sticker here.

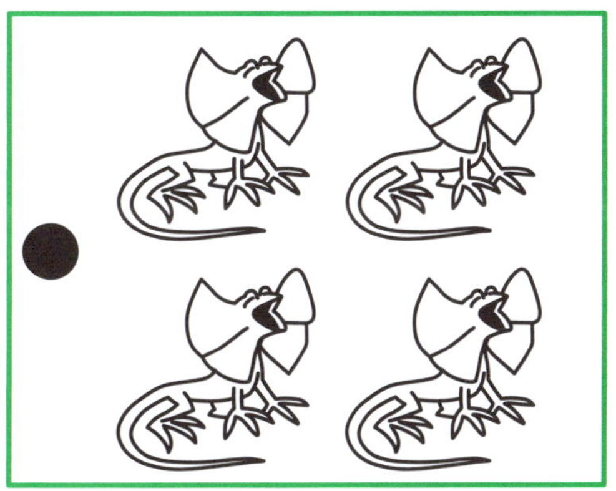

2. Ask your child to colour in each group of 5 on this page in the same coloured crayon.

Place a sticker here.

Finding the matching number

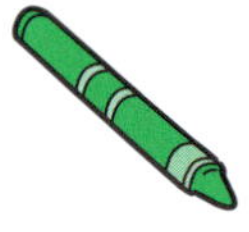

How many people or animals can you see in each box? Circle the matching number.

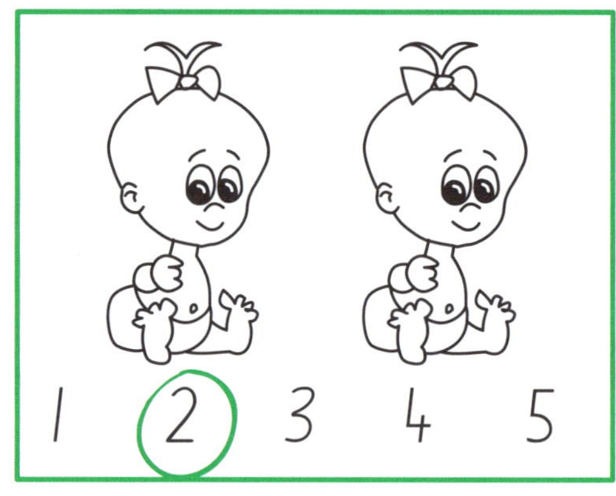

1. Ask your child to count all the smiles on this page. How far can they count past 5?

Place a sticker here.

from 1–5

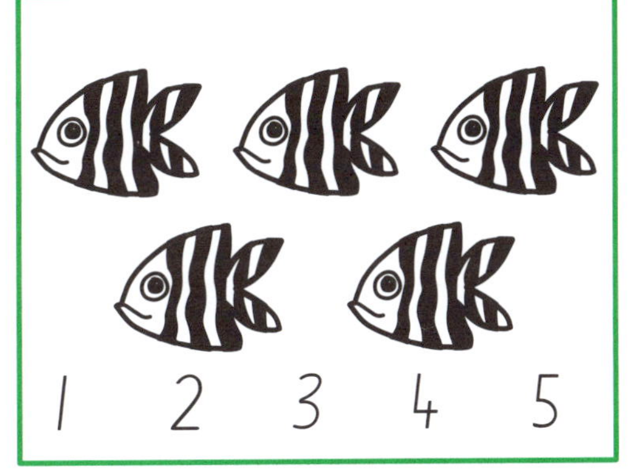

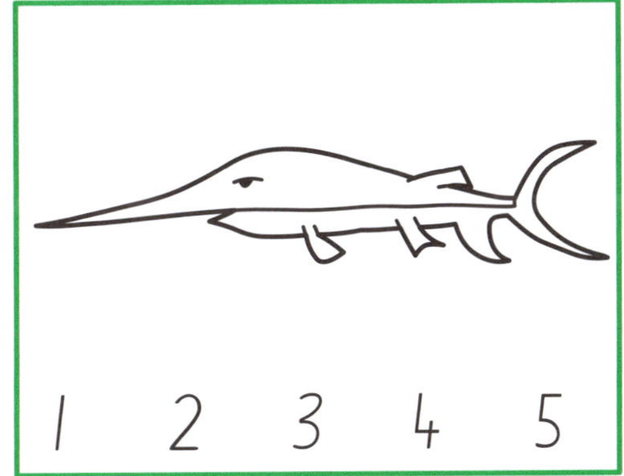

2. Ask your child if there are more jellyfish than sharks on this page. Then count them to check.

Place a sticker here.

Drawing matching objects

What's your favourite biscuit? What's your favourite fruit? Draw biscuits to match the number on each jar. Then draw fruit to match the numbers next to each basket.

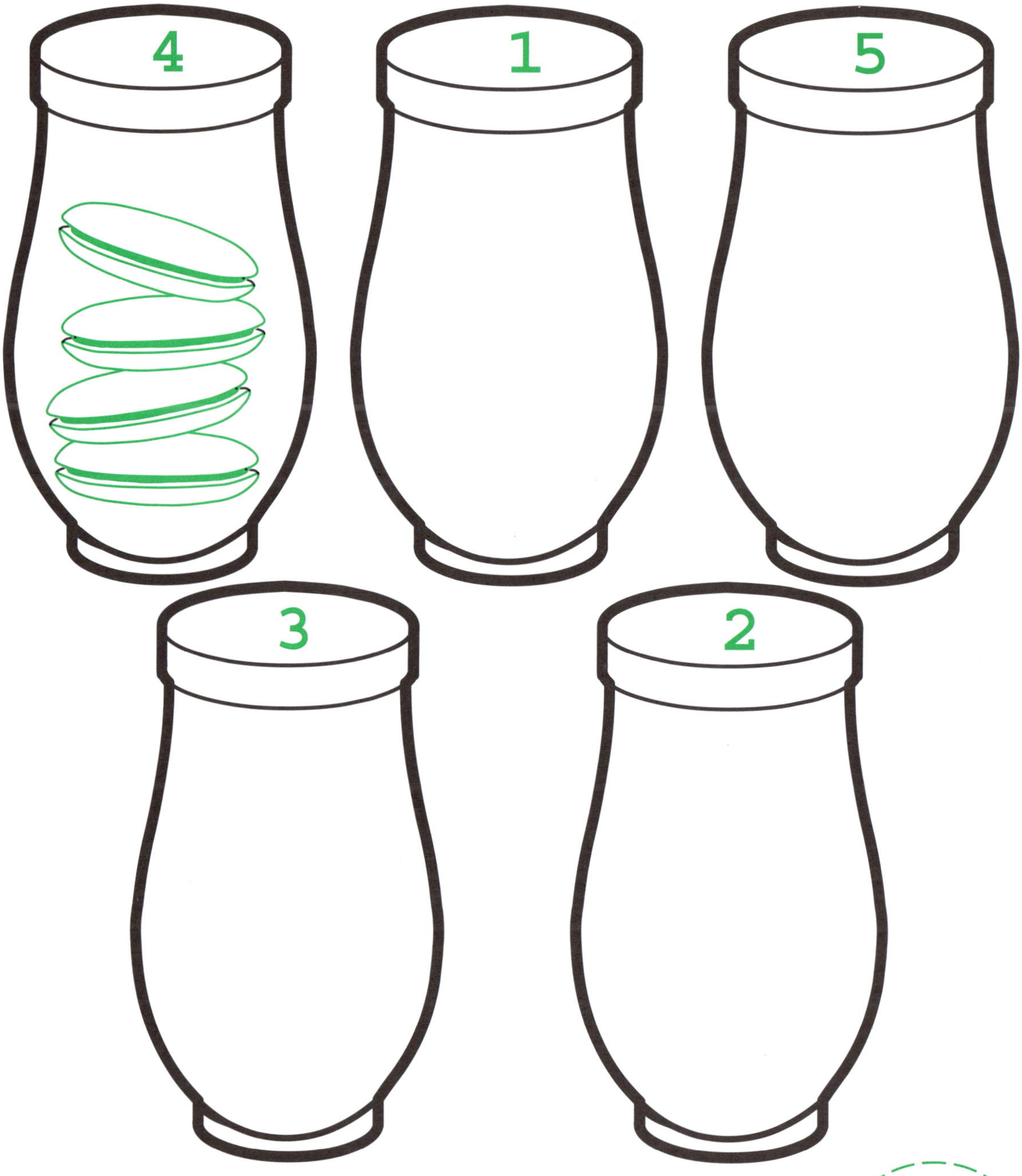

4 1 5

3 2

1. How many biscuit jars are on this page? Ask your child to count them.

Place a sticker here.

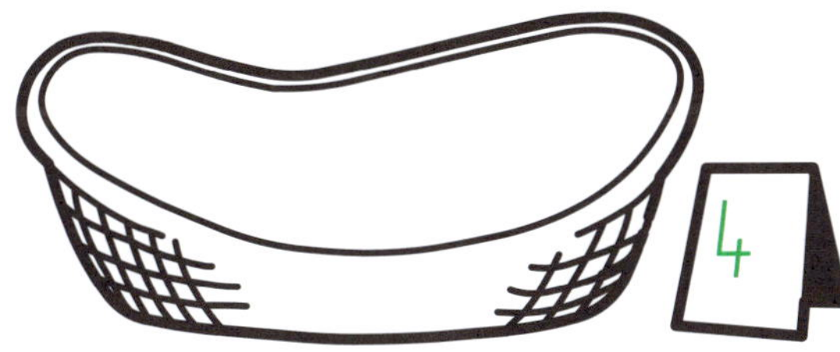

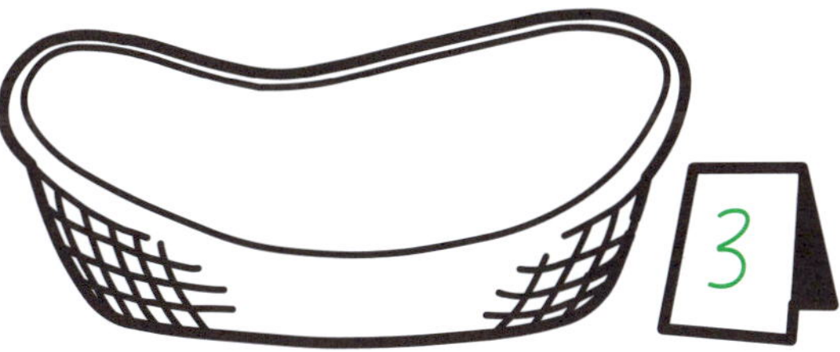

2. How many baskets are on this page? Ask your child to count them.

Place a sticker here.

Writing the matching number

How many dots are on the dice? How many spots are on each beetle? Count them, then write the matching number beside each picture.

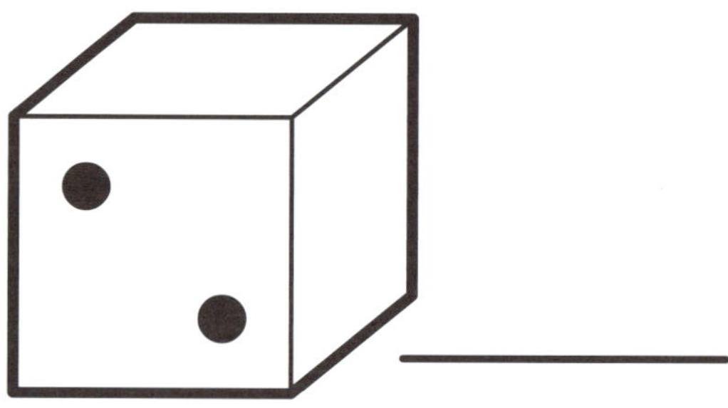

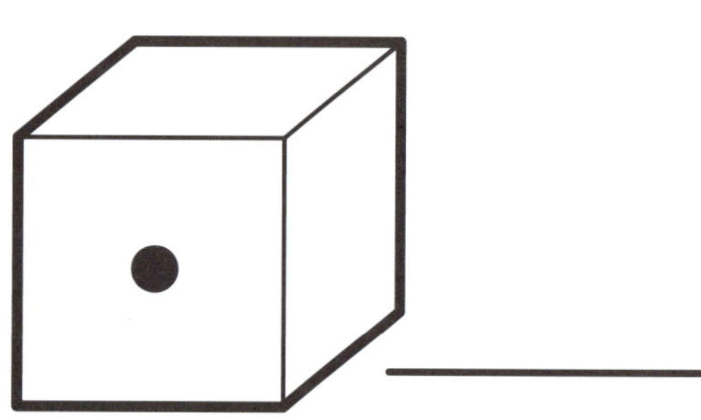

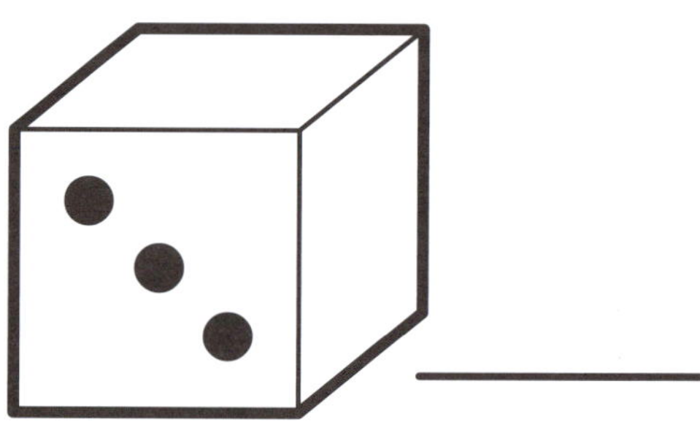

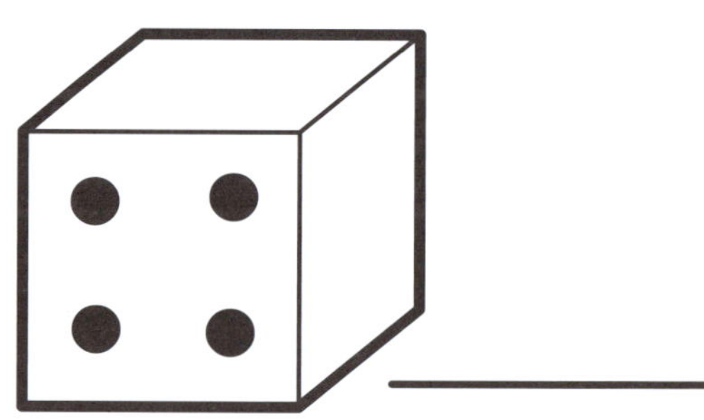

1. If your child enjoys this activity, you could try rolling real dice.

Place a sticker here.

from 1–5

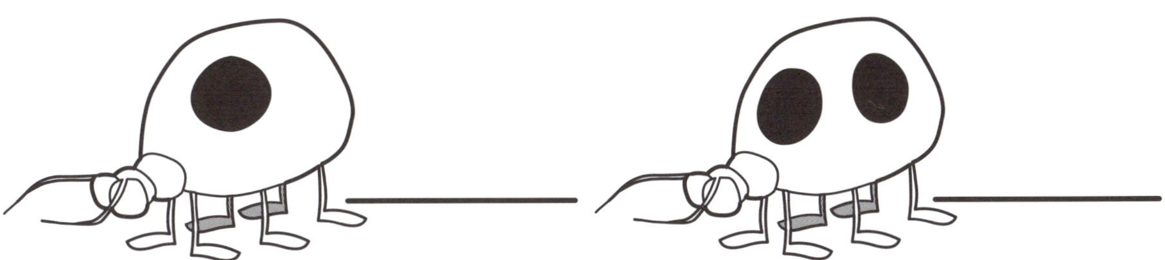

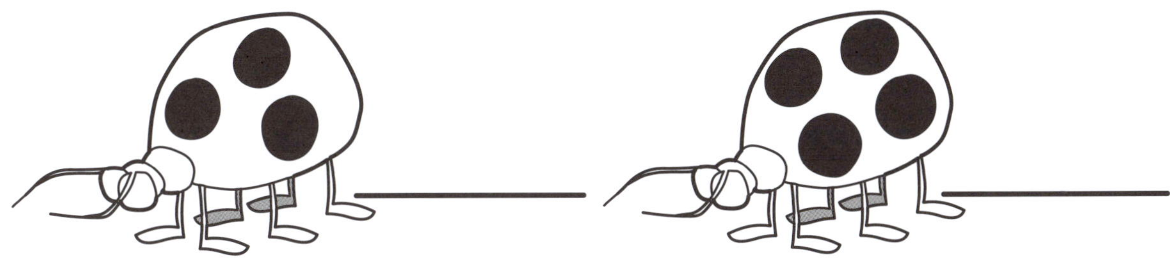

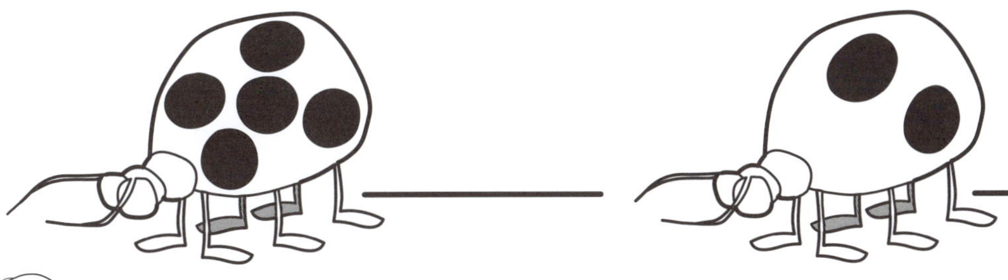

2. Ask your child to count how many beetles have 5 spots.

Place a sticker here.

Colouring by numbers

Pick five different coloured crayons. Colour in each triangle at the top of the page in a different colour, then colour the pictures to match.

1 △ 2 △ 3 △ 4 △ 5 △

3

2 4

2

1 3 5

1 1 5

3 3

4 1 1

5

2

4 4

1. Ask your child to find each number from 1 to 5 around the house.

Place a sticker here.

1 2 3 4 5

1

2

1

5 5

3

2 2

4

5 5

2. You can make an activity like this with any colouring book picture. Add numbers to the pictures, then draw up a colour code at the top of the page.

Place a sticker here.

Well done!

You have finished the book !

Place your last two stickers on the picture.
You can colour in the picture, too.